Compost Made Easy

T.S. AVINI

Compost Made Easy

Recycle Waste and Feed Your Garden Naturally

by T.S. AVINI

Table of Contents

Chapter 1: Introduction to Composting: A Natural Recycling

In a world where sustainability is becoming increasingly important, composting stands out as a simple yet highly effective way to contribute to environmental preservation.

Composting is more than just a gardening habit; it is a form of natural recycling that harnesses the processes of nature to transform organic waste into nutrient-rich soil. By integrating composting into your life, you can reduce waste in landfills, lower your carbon footprint, and enrich the ecosystem of your own garden.

The Marvel of Natural Processes

At first glance, Before diving into the practical aspects of composting, it is essential to understand the natural processes that make it possible. Composting is the decomposition of organic matter into a dark, crumbly, earthy-smelling material known as compost.

This transformation occurs through the collective action of a diverse group of microorganisms, such as bacteria and fungi, as well as larger organisms, including earthworms and insects. These organisms feed on organic waste, breaking it down into simpler components. The result is a material rich in essential nutrients like nitrogen, phosphorus, and potassium—key elements required for healthy plant growth.

Unlike synthetic fertilizers, which can leach chemicals into the soil and water, compost serves as a harmless and balanced amendment to the earth.

When added to the garden, compost improves soil structure, encourages the presence of beneficial organisms, and enhances water retention. It's nature's way of giving back, closing the loop on waste by recycling nutrients back into the land.

Why Compost?

Many people don't realize this, but Many of us contribute to an incredible amount of waste without even realizing it. Household waste primarily comprises materials that are highly compostable, such as fruit and vegetable peels, coffee grounds, eggshells, yard clippings, and shredded paper.

By choosing to compost, we reduce the volume of waste directed to landfills, which are already overburdened. Decomposing organic matter in landfills emits methane, a potent greenhouse gas that contributes significantly to climate change. In contrast, composting converts this potential environmental harm into a resource for a sustainable future.

Moreover, composting can also lead to financial savings.

By reducing reliance on commercial fertilizers and soil conditioners, you lower your gardening costs. Additionally, as communities adopt composting programs, municipalities can benefit from reduced waste management expenses.

Debunking Composting Myths

Interestingly enough, One of the hurdles to getting started with composting is dispelling the misconceptions that often accompany it.

Many believe that composting is a time-consuming or difficult process. Indeed, while it does require some attention and patience, it is not nearly as challenging as commonly perceived.

There is also a fear that composting might produce foul odors or attract pests.

However, properly managed compost piles should exude a pleasant, earthy aroma, and pests can usually be avoided by maintaining the right balance of green (nitrogen-rich) and brown (carbon-rich) materials and avoiding inappropriate waste like meat and dairy products.

Starting Your Compost Journey

In real life, Beginning your compost journey involves a few simple considerations: space, materials, and methodology.

You don't need acres of land to start composting—a small backyard, a corner of a balcony, or even a composting bin in a kitchen can suffice. The key is to find a system that suits your lifestyle and space constraints.

Types of Composting Systems

There are various composting systems to accommodate different living situations and gardening needs:

It's worth noting that 1.

Traditional Compost Pile: For those with outdoor space, this system is the most straightforward. It involves simply piling your compostable materials in a designated area and letting nature do its work.

2.

Compost Bins: These come in a range of designs and sizes, offering a more contained and sometimes more visually appealing option than open piles. Tumbling composters make it easy to aerate your pile by turning it regularly.

3.

Vermicomposting: Ideal for small spaces or indoor composting, vermicomposting uses worms, often red wigglers, to accelerate the decomposition process. The result is worm castings, a highly potent fertilizer.

4.

Bokashi Composting: Originating from Japan, this method involves fermenting organic waste using a special inoculated bran. The Bokashi process is compact and can handle dairy and meat without the pests or odors.

Building Your First Compost Pile

To build a successful compost pile, you'll need a balanced mix of green and brown materials. Green materials include fruit and vegetable scraps, coffee grounds, and grass clippings.

These provide nitrogen. Brown materials, like dead leaves, straw, and newspaper, offer carbon. Striking the right balance—often suggested as two-thirds brown to one-third green—ensures the pile remains aerated and free from odor.

At first glance, Layering is another important practice.

Start your pile with coarse materials to help with aeration and then create alternating layers of green and brown materials. Moisture is key—your compost pile should feel like a well-wrung sponge, moist but not soggy.

Aeration promotes aerobic digestion by the microbes, preventing the pile from becoming anaerobic and smelly.

This can be achieved naturally with the right mix of materials or by turning the compost regularly with a pitchfork.

Looking Ahead

Let's not forget, As you embark on your composting journey, patience and observation will be your greatest allies. Composting is a dynamic process and will adapt to the changes in materials and conditions you provide.

By learning to balance moisture, material ratios, and aeration, you will fine-tune your composting skills. And over time, you will witness the transformation of waste into a valuable asset for your garden—a testament to the restorative power of natural recycling.

With a solid understanding of its benefits and processes, you are well on your way to joining the ranks of conscientious gardeners and environmental stewards transforming their organic waste into rich compost.

In doing so, you follow the example set by nature, where no waste is left behind, and everything has its place in the cycle of life.

Chapter 2: The Science Behind Compost: How it Works

In our quest to create a thriving, sustainable garden, understanding the scientific processes that underpin composting is essential.

Composting is a natural process that transforms organic matter into a valuable soil amendment. In this chapter, we'll delve into the intricacies of how composting works, unraveling the biological, chemical, and physical processes that transform kitchen scraps and yard waste into nutrient-rich compost.

The Soil Food Web: A Dynamic Ecosystem

To appreciate how compost works, we must first explore the concept of the soil food web.

The soil food web is a complex system of interactions between microorganisms, insects, worms, and other organisms in the soil. Within this web, each organism plays a role in breaking down organic material.

At the core of composting are microorganisms such as bacteria and fungi. These tiny, yet powerful agents of decomposition are the primary drivers of the composting process.

They consume organic matter, breaking it down into simpler compounds in a process known as biodegradation. This breakdown releases nutrients and creates the rich humus-like substance that we call compost.

The Role of Microorganisms

Let's not forget, Bacteria are the first responders in the compost pile. These single-celled organisms are incredibly diverse and capable of breaking down a wide array of organic materials.

They thrive in the warm, moist conditions of a compost pile and are responsible for the majority of initial decomposition. Two main types of bacteria are involved: mesophilic and thermophilic bacteria.

Mesophilic bacteria function at moderate temperatures, typically between 70°F and 100°F. They start the composting process by rapidly consuming sugars and starches, generating heat in the process.

As the pile heats up, **thermophilic bacteria** take over. These heat-loving microbes thrive between 104°F and 160°F and continue the decomposition of tougher plant fibers and proteins.

Fungi, particularly molds and yeasts, are equally important in composting.

Fungi excel at breaking down complex organic compounds like lignin, which makes up the woody parts of plants. These organisms work more slowly than bacteria but play a vital role in decomposing materials that other microbes find challenging to digest.

The Chemical Transformations

The decomposition of organic matter in a compost pile involves a series of chemical reactions.

The primary chemical transformations encompass the conversion of complex organic compounds into simpler molecules.

Carbon and Nitrogen Ratio (C:N Ratio): One of the most critical factors in composting is maintaining a balanced carbon-to-nitrogen ratio. Carbon provides energy for microbes, while nitrogen is necessary for protein synthesis.

An ideal C:N ratio for a compost pile is about 25-30:1. Materials high in carbon include leaves, straw, and paper, while those high in nitrogen are typically food scraps and grass clippings. Balancing these elements ensures efficient decomposition and minimizes unpleasant odors.

I've seen this happen before: **Oxygen:** Composting is an aerobic process, meaning it requires oxygen.

Oxygen supports the metabolic functions of aerobic microorganisms. Regularly turning or aerating the compost pile introduces fresh air, preventing the pile from becoming anaerobic (lacking oxygen), which can lead to the production of foul-smelling gases.

Moisture: A compost pile should be as moist as a wrung-out sponge.

Water acts as a solvent, facilitating the movement of nutrients and allowing microorganisms to thrive. Too little moisture slows down microbial activity, while excess moisture displaces air, leading to anaerobic conditions.

The Physical Structure of Compost

I've seen this happen before: The physical structure of a compost pile influences its effectiveness and efficiency.

A well-structured pile provides the optimal conditions for microbial activity.

The ideal compost pile has a mix of particle sizes to ensure adequate airflow.

Larger particles like twigs and straw create air pockets, while smaller materials like grass clippings fill in gaps. This structure helps maintain oxygen levels and encourages the even distribution of moisture and microorganisms.

Layering is another crucial technique.

Alternate layers of carbon-rich and nitrogen-rich materials, beginning with coarse materials at the bottom, create a balanced environment. This approach mimics the natural layering of debris on a forest floor, promoting efficient decomposition.

The Breakdown Process: From Waste to Wealth

Now that we've examined the components of composting, let's explore the stages of the breakdown process:

1. **Mesophilic Phase:** During this initial phase, mesophilic bacteria begin to decompose simple carbohydrates and proteins. The pile gradually heats up as microbial populations grow.

It's worth noting that 2.

Thermophilic Phase: As temperatures rise, thermophilic bacteria take over, breaking down more complex compounds. This intense microbial activity can raise the temperature of the compost pile to over 140°F, effectively killing weed seeds and pathogens.

3. **Cooling and Curing Phase:** As easily digestible materials become scarce, microbial activity decreases, and the pile begins to cool.

Fungi play a more prominent role, breaking down tougher materials. The compost stabilizes, transitioning from an active heap to a mature blend.

4.

Finished Compost: Finally, your compost is ready for use when it appears dark, crumbly, and earth-like, with a pleasant, earthy smell. This nutrient-rich mixture can be added to your garden, enhancing soil structure, moisture retention, and nutrient availability.

Creating Your Ideal Compost Environment

To create the most effective compost, start with a variety of materials to ensure a balanced C:N ratio.

Regularly turn the pile to maintain oxygen levels, and monitor moisture content closely. While most compost piles reach temperatures sufficient to decompose organic matter, insulated bins or covers can help retain heat in cooler climates.

Remember that patience is key.

Composting is a natural process that takes time, but the benefits are well worth the wait. By understanding the science behind composting, you can optimize your efforts and produce a consistent supply of high-quality compost for your garden.

I've seen this happen before: In the next chapter, we'll explore the essential tools and materials you'll need to get started on your composting journey.

With the knowledge you've gained here, you're well on your way to becoming a master composter, transforming waste into a valuable resource that nurtures your garden naturally.

Chapter 3: Understanding Different Composting Methods

In the world of composting, there isn't a one-size-fits-all solution. Different environments, time constraints, and even personal preferences make certain composting methods more suitable than others.

To become a master composter, it's crucial to understand the various methods available and determine which one aligns best with your lifestyle and gardening goals. In this chapter, we'll explore the most popular composting techniques, discussing their advantages, drawbacks, and ideal conditions for usage.

Traditional Hot Composting

Let's not forget, Hot composting is perhaps the most common and widely practiced method.

It involves actively managing a compost pile to maintain high temperatures, which accelerate the breakdown of organic material and kill pathogens and weed seeds.

How It Works

To establish a hot compost pile, start by gathering a balanced mix of green (nitrogen-rich) materials like grass clippings and kitchen scraps, and brown (carbon-rich) materials such as dry leaves and straw. Aim for a carbon to nitrogen ratio of approximately 30:1.

Layer these materials in a pile or bin, ensuring each layer is slightly moist. To achieve the required high temperatures (between 130-160°F), the pile needs to be at least 3 feet in height and width.

Practical Tips:

- **Turning:** Regularly turning the pile every few weeks introduces oxygen, a critical element for aerobic decomposition.

- **Moisture Control:** The pile should feel like a damp sponge – not too dry but not waterlogged.

In real life, - **Monitoring:** Using a compost thermometer can help you manage the temperature effectively.

Advantages:

- Faster decomposition, usually within 1 to 3 months.

- Kills most pathogens and weed seeds.

Disadvantages:

- Requires more effort and attention.

- Space-intensive and might not be suitable for smaller gardens.

Cold Composting

I've seen this happen before: Cold composting is the slower, hands-off sibling of hot composting.

This method requires minimal effort but takes longer to yield usable compost.

How It Works

Simply add organic materials to a heap as they become available. There is no need to worry about the carbon-to-nitrogen ratio or temperature.

Over time, the materials will decompose on their own. This process can take anywhere from 6 months to 2 years.

Advantages:

- Minimal maintenance is required.

- Suitable for those with limited time or interest in active management.

Disadvantages:

- Longer decomposition time.

Many people don't realize this, but - Weeds and seeds might survive the process, requiring careful sifting before use.

Vermicomposting

For those interested in a more interactive and contained composting method, vermicomposting employs worms to break down organic material.

Red wigglers (Eisenia fetida) are the most common worm species used for this purpose.

How It Works

Vermicomposting systems are typically housed in bins. Start by adding bedding material such as shredded newspaper or cardboard.

Introduce the worms and gradually feed them small amounts of kitchen scraps. It's important to avoid adding too much food at once, as this can lead to odors and attract pests.

Practical Tips:

- **Bin Conditions:** Keep the bin in a cool, dark place, ideally between 55 and 77°F.

I've seen this happen before: - **Avoid:** Citrus, onions, and meat, which can disrupt the ecosystem and harm the worms.

- **Harvesting:** Worm castings, or vermicompost, can be harvested every few months for nutrient-rich garden soil.

Advantages:

- Efficient in small spaces, making it ideal for apartments.

- Produces high-quality compost, rich in nutrients and beneficial microbes.

Disadvantages:

- Requires ongoing care and monitoring.

Let's not forget, - Worms need to be protected from extreme temperatures and excess moisture.

Bokashi Composting

Bokashi is a fermentation-based method originating from Japan, particularly useful for composting meat, dairy, and other difficult materials that might not work well in traditional composting.

How It Works

Bokashi relies on a specialized bran inoculated with effective microorganisms (EM) that ferment organic waste in an airtight container. Kitchen scraps are layered with the bran and pressed down to remove air.

Once the container is full, seal it and allow it to ferment for two weeks. The fermented material can then be buried in soil to complete the breakdown.

Advantages:

- Handles a wider range of materials, including meats and bones.

- Quick process, with initial fermentation taking only a couple of weeks.

Disadvantages:

Believe it or not, - Requires purchase of Bokashi bran and special airtight bins.

- Fermented material still needs to be disposed of properly by burying it in soil.

Trench Composting

For those with space in their garden, trench composting offers a simple and unobtrusive method.

This technique composts material directly into the soil, feeding plants at the root level.

How It Works

It's worth noting that Dig a trench about a foot deep in an unused garden area.

Add kitchen waste and other compostables, then cover with soil. Over a few months, the materials will decompose, enriching the soil directly.

Advantages:

- No turning or active management necessary.

- Enriches the soil in its natural location.

Disadvantages:

- Requires available garden space.

- Takes several months to be effective, thus not ideal for immediate planting.

Conclusion

Believe it or not, Choosing the right composting method depends on various factors, including space, time, the type of waste you regularly produce, and personal preferences. While hot composting suits the dedicated gardener, vermicomposting might appeal to someone in an urban setting.

On the other hand, Bokashi lets those dealing with meat and dairy participate effectively in composting. And for those seeking simplicity, trench composting offers an easy, low-maintenance option. Understanding these methods gives you a leg up in making composting a seamless part of your everyday life, ensuring a sustainable and productive approach to gardening.

Chapter 4: Essential Tools and Equipment for Composting

In the world of composting, having the right tools and equipment can mean the difference between a smooth, efficient process and a frustrating experience laden with obstacles.

Many new composters underestimate the value of investing in quality tools, but as your composting journey progresses, the rewards of being properly equipped become apparent. This chapter will guide you through the essential tools and equipment needed for effective composting, ensuring you are prepared to turn your kitchen and yard waste into nutrient-rich compost.

Compost Bins: Choosing the Right Home for Your Compost

The first decision any novice composter faces is selecting an appropriate compost bin.

Bins keep your compost pile neat, help retain moisture, and keep pests out. While technically, you can compost in an open pile, bins provide benefits that often make the investment worthwhile.

There are several options available, from simple DIY setups to commercially manufactured bins:

1.

Stationary Bins: Ideal for those with more space and who prefer a 'set-it-and-forget-it' approach. These bins are typically affordable, easy to use, and vary in size.

Let's not forget, 2.

Tumbler Bins: Perfect for the more committed composter looking for faster results. A tumbler helps you aerate the compost by rotating it with minimal effort, speeding up the decomposition process.

3.

DIY Bins: If you're handy and enjoy a project, building your own bin using materials like wire mesh or wooden pallets can be cost-effective and customizable to your space.

4.

Worm Bins: For those embracing vermicomposting, worm bins are essential. These bins, often smaller and suited for indoor use, host worms that break down waste even more efficiently.

Aeration Tools: Keeping It Breathable

Oxygen is a critical element in the composting process, as it fosters the health of the aerobic bacteria responsible for decomposition.

Regularly aerating your compost heap ensures it remains oxygenated and avoids becoming anaerobic, which can lead to unwanted odors:.

1.

Compost Aerator: This tool is explicitly designed to penetrate the compost pile and create channels for airflow. It generally consists of a long, stick-like tool with blades or wings at the bottom that churn and fluff the pile when plunged in.

I've seen this happen before: 2.

Garden Fork: A regular garden fork or pitchfork can suffice for turning larger compost heaps manually. It is a versatile tool that can also be used for other gardening activities.

3. **Hole-Puncher or Aerating Tool for Tumblers**: Some tumbler bins include optional aerating accessories that can improve airflow without the need for extensive turning.

Moisture Tools: Managing Water Levels

Believe it or not, Moisture balance is crucial in composting.

Too little moisture will slow down decomposition, while too much can drown the microbes and lead to a smelly pile:.

1.

Garden Hose or Watering Can: To adjust the moisture level, a source of water is necessary. A watering can is efficient for smaller bins, while a garden hose with adjustable settings is ideal for larger piles.

2.

Compost Thermometer: Though not directly moisture-related, a thermometer measures the internal temperature of the compost, indirectly allowing you to infer moisture levels. A damp but not soggy pile that generates heat indicates the correct moisture.

It's worth noting that 3.

Tarp or Cover: Protecting your compost bin from excessive rain or dry heat is essential. A tarp or custom bin cover keeps the elements in check.

Cutting and Chopping Tools: Preparing Your Materials

To create the optimal environment for decomposition, materials should ideally be small and of uniform size:

It's worth noting that 1.

Garden Shears or Secateurs: These will help trim larger plant material or shrub prunings down to a manageable size. Shears are particularly useful for breaking down stubborn garden waste into smaller pieces.

2.

Chipper or Shredder: While more of an investment, a shredder or chipper can handle tougher materials like branches and make quick work of reducing them to compost-friendly pieces, accelerating the breakdown process.

Collection and Transport Tools: Managing Your Waste

Let's not forget, Efficiently managing the influx of materials destined for your compost bin involves having the right tools for collecting and transporting:

1. **Kitchen Compost Caddy**: Keep this small, sealed bin under your sink for kitchen scraps. Opt for one with a charcoal filter to manage odors.

2.

Wheelbarrow or Garden Cart: If you have a larger garden or yard, transporting waste material with a wheelbarrow or garden cart becomes essential. It cuts down on time and effort needed compared to hand-carrying materials.

Let's not forget, 3.

Compost Sifter: Once your compost is ready, a sifter can help refine the final product by removing large, undecomposed chunks, resulting in a finer compost ideal for mixing into soil.

Personal Safety and Maintenance: Ensuring Longevity and Health

Your safety and the longevity of your tools are equally important.

I've seen this happen before: 1. **Gloves**: Protect your hands from sharp objects, insects, and possibly harmful bacteria with durable gardening gloves.

2.

Tool Maintenance Kit: Keep a basic kit with lubricants, sharpening tools, and cleaning supplies to ensure your tools stay in top working condition. Regular maintenance will prolong the life of your equipment and save you replacement costs in the long run.

Equipping yourself with these essential tools and equipment sets a solid foundation for successful composting. By making wise investments early in your composting journey, you'll maximize your efficiency and soon enjoy the satisfaction of producing your own rich, earth-friendly compost.

As you progress, you may find your needs evolving, and adding to your collection of tools can make composting even easier and more rewarding. Whether your goal is to reduce landfill waste, improve your garden's fertility, or simply engage in a sustainable practice, having the right tools on hand makes all the difference in achieving smooth and effective composting.

Chapter 5: Starting Your Compost: Selecting the Right Bin

Venturing into the rewarding world of composting begins with one crucial decision: selecting the right compost bin.

With a myriad of options available, the choice may seem daunting. However, by understanding your specific needs, existing garden space, and personal preferences, you can make a choice that simplifies the composting process and enhances your garden's health.

Understanding the Basics

Many people don't realize this, but Before diving into the various types of compost bins, it's important to first understand what a compost bin must accomplish. Essentially, a compost bin's primary function is to house organic materials while providing the right conditions for them to decompose.

This means containing waste, maintaining moisture, allowing for aeration, and keeping out pests. Every bin, no matter its design, should address these basic needs.

Types of Compost Bins

1. **Open Pile or Heap**

I've seen this happen before: If you have ample space and aren't overly concerned about aesthetics, an open pile is the simplest method. This option involves designating a space directly on the ground for organic waste.

It benefits from unrestricted air flow, but requires regular turning to aerate the core materials. A major downside is its vulnerability to pests like rats and raccoons. To mitigate this risk, some gardeners choose to encircle their heap with chicken wire or mesh.

2. **Stationary Bins**

Stationary bins are a popular choice in urban and suburban areas due to their compact nature. They are usually made from wood, plastic, or metal, and range in size and shape.

A stationary bin allows for neat and controlled composting, making it difficult for pests to access. They come either as DIY designs—offering the satisfaction of customization—or as readymade solutions from garden centers. Look for bins with airflow features, such as ventilated sides or lids, to ensure decomposition is efficient.

3. **Tumbling Composters**

For those who prefer minimal labor, tumbling composters are a fantastic innovation. These bins are designed for easy turning; the barrel rests on a stand allowing you to rotate it with minimal effort.

This feature is particularly helpful in ensuring even decomposition without manually turning the pile with a pitchfork. Tumblers are typically off the ground, which reduces the risk of pests. Bear in mind, however, that most tumblers have a limited capacity and might not suffice for those with extensive garden waste.

4. **Worm Bins**

As strange as it sounds, For indoor or small-space composting, consider a worm bin, or vermicomposter. These bins utilize red worms to break down organic matter at impressive speeds, producing nutrient-rich castings.

They are compact and can fit in kitchens or balconies, making them an ideal choice for apartment dwellers. While requiring a bit more care (maintaining correct moisture levels and feeding the worms properly), vermicomposting offers an incredible payoff in the form of rich, balanced compost.

5. **Anaerobic Bins**

Anaerobic bins, or Bokashi bins, offer a unique way of composting using a fermentative process. These bins require the addition of specialized microorganisms, typically in the form of a bran treated with bacteria.

The process operates without air and is notably faster, suitable for those who want quick results. However, the process also produces a liquid byproduct that must be managed regularly.

Factors to Consider

Choosing the right bin involves weighing several practical factors:

- **Space Availability**: Assess the area where the bin will be placed.

Will it fit within your garden layout, on your balcony, or even under the kitchen sink? Larger areas may accommodate expansive piles or multiple compost stations, while confined spaces require more compact systems.

Many people don't realize this, but - **Volume of Waste**: Consider the amount of organic waste you generate.

Those with large gardens or who maintain vegetarian households might need larger bins or multiple units. Smaller households might find a single, compact bin sufficient.

- **Effort and Maintenance**: Your willingness to engage affects the choice significantly.

Open piles require frequent turning, while tumblers simplify this task. Similarly, knowing whether you can commit to maintaining the moisture and feed for a worm bin can influence your decision.

- **Pest Management**: In areas prone to pest problems, secure bins with sturdy lids, and off-ground tumbling systems are advantageous.

Let's not forget, - **Aesthetics and Regulations**: Lastly, consider how your composting setup blends with the surroundings and adheres to any local residential guidelines.

In urban settings, well-designed, aesthetically pleasing bins can be particularly important.

Making the Bin Work for You

Once chosen, maximizing your compost bin's effectiveness is key. Start by situating the bin optimally; it should be on level ground with a blend of sun and shade to regulate temperature and moisture.

Fill it with a balanced mix of green and brown materials, and water sufficiently to keep the composite damp but not soggy. Avoid materials that can attract pests or cause odors, such as dairy and meat. Turn the contents periodically unless using a stationary setup to promote even breakdown.

Conclusion

Selecting the right compost bin is a foundational step that influences the efficiency and enjoyability of your composting journey.

By considering the space you have, the waste you produce, and the level of effort you're willing to invest, you can choose a bin that aligns perfectly with your needs. As you embark on this eco-friendly journey, remember that the right bin doesn't just contain waste—it transforms it into the lifeblood of a thriving, sustainable garden.

Chapter 6: Choosing the Best Location for Your Compost

Once you have decided to embark on the rewarding journey of composting, one of your first crucial decisions is selecting the ideal location for your compost heap or bin. Just as with real estate, when it comes to composting, it's all about location, location, location.

The spot you choose will not only affect the efficiency of your composting process but also its convenience and overall success. In this chapter, we'll explore the key factors to consider to help you pick the perfect spot to transform kitchen scraps and yard waste into rich, black gold for your garden.

Understanding Composting Needs

Let's not forget, Before settling on a location, it's essential to understand what compost requires to decompose effectively and efficiently.

Your compost heap relies on a balance of four main elements: greens (nitrogen-rich materials), browns (carbon-rich materials), moisture, and air. Therefore, the site should support these needs by allowing for adequate aeration, drainage, and accessibility to the elements.

Sunlight Exposure: Compost piles thrive with occasional sun exposure, as the warmth accelerates the decomposition process.

However, too much direct sunlight will dry the pile out, necessitating frequent watering, especially during hotter months. The ideal location will have dappled sunlight or partial shade, ensuring the pile doesn't dry out or become waterlogged.

Proximity and Convenience

An often-overlooked aspect of compost site selection is its convenience. You'll likely be making frequent trips to your compost bin—after all, it's meant to dispose of kitchen scraps, yard waste, and other organic materials regularly.

Select a spot that is easily accessible from your kitchen and your garden, so adding materials and retrieving compost later is a breeze. When your compost site is inconvenient, it's easy to lapse into neglect, which can halt your composting efforts.

Drainage and Foundation

Appropriate drainage is crucial for efficient composting because too much water can lead to odors and sluggish decomposition, while too little can cause dryness and slow down the process.

Avoid low-lying areas where water collects after rain, as these can become boggy and make it difficult for the compost heap to maintain the right balance.

Believe it or not, The foundation of your compost pile should ideally be on bare earth.

This setup allows beneficial organisms—like worms and microbes—to migrate into the pile, adding to the decomposition process. However, if you are using a bin with a solid base, ensure there is sufficient drainage and airflow to prevent water from accumulating.

Odor and Aesthetics

Even a well-maintained compost pile can produce off-putting smells periodically, especially when an imbalance occurs. To minimize potential smells, place your compost area downwind of your home and outdoor living spaces.

Additionally, consider the aesthetic aspect—while some compost bins are designed to be visually appealing, a heap of decomposing leaves and grass might not blend well with your garden's visual appeal. Strategically place the compost area so that it is functional but does not detract from the enjoyment of your outdoor space.

Local Regulations and Neighbor Considerations

Before finalizing your compost location, familiarize yourself with any local regulations about composting.

Some urban or suburban areas may have specific guidelines or restrictions that dictate where compost can be placed, how it should be enclosed, or even the size of the compost heap. By adhering to these guidelines, you avoid potential fines or complaints from neighbors.

Speaking of neighbors, it's courteous to inform them of your composting plans if your property is close to theirs.

This proactive communication can help alleviate any concerns they might have about smells or pests.

Pest Prevention

Pests such as rodents or raccoons can be an issue if your compost is easily accessible to them. Situating the compost pile away from fences or structures that could provide shelter for pests is wise.

Furthermore, if you live in an area where wildlife is a concern, consider an enclosed compost bin. This design not only deters pests but also keeps your compost neatly managed and out of sight.

Wind Exposure

Interestingly enough, Strong winds can excessively dry out your compost heap and even blow away lighter materials.

Choosing a site with a natural windbreak—such as a hedge, wall, or sturdy fence—can help maintain the pile's moisture balance while still allowing for adequate airflow. If natural barriers aren't available, consider creating a windbreak with portable fencing or by planting fast-growing shrubbery.

Seasonal Considerations

Think about how different seasons might affect your compost location. An area that seems perfect in the summer may become less ideal in the winter if it becomes inaccessible or difficult to manage because of snow or mud.

Likewise, a spot that is too shady in winter may become too sunny in the summer. Ideally, choose a location that remains functional year-round, taking into account the changes each season brings.

Test and Adjust

After selecting a location and setting up your compost heap, monitor its decomposition progress and adjust as necessary.

It takes time to achieve the ideal balance, so don't be discouraged if the location needs tweaking. Continually observe how sun, shade, and other environmental factors impact the process to fine-tune your compost site over time.

In conclusion, selecting the best location for your compost is about finding balance.

By taking into account sunlight, convenience, drainage, aesthetics, and other factors, you can create an inviting environment that will support efficient decomposition. This careful consideration will result in a composting system that is practical, effective, and rewarding, providing your garden with the nutrients it needs to flourish naturally.

Chapter 7: What to Compost: A Guide to Acceptable Materials

Composting is a cycle that mimics the natural processes of decomposition, turning organic waste into rich, nutrient-dense fertilizer.

To become a proficient composter, it is essential to understand what materials can be composted. This chapter will walk you through the types of materials that are suitable for composting, categorizing them into 'greens' and 'browns' to help you maintain a balanced and effective compost pile.

The Basics of Browns and Greens

Let's not forget, Composting materials are typically categorized into 'greens' and 'browns'.

Understanding the distinction between these two types is crucial for maintaining the right balance in your compost pile.

1. **Browns**: These materials are rich in carbon and are usually dry and woody.

They provide the necessary air pockets that help keep the compost pile aerated. Some common browns include:.

In real life, - **Dried leaves**: An abundant resource, especially in autumn. They break down slowly but add essential carbon to your mix.

- **Straw and hay**: These are great for aeration and also contribute carbon. Be cautious of hay as it can contain seeds, which might germinate.

- **Cardboard and paper (uncoated)**: Shredded paper and brown cardboard, such as paper towel rolls and boxes, are excellent in moderation.

Avoid glossy or coated papers.

As strange as it sounds, - **Wood chips and sawdust**: Perfect for carbon but use wood products in moderation since they break down slowly.

2. **Greens**: Rich in nitrogen, greens are the moist, fresh organic matter crucial for feeding the microorganisms that decompose your compost. Key examples include:

In real life, - **Vegetable and fruit scraps**: Everything from apple cores to carrot peels. Avoid excessive citrus peels as they can be too acidic.

- **Coffee grounds and tea bags**: Although dark, coffee grounds are considered greens and are a nitrogen-rich addition.

Ensure tea bags are compostable, as some contain synthetic fibers.

- **Fresh grass clippings**: These add abundant nitrogen, but in large quantities, they can compact and inhibit airflow. Mix with browns to avoid this.

Let's not forget, - **Eggshells**: Technically not a green, but as a calcium source, crushed eggshells are valuable, though they decompose slowly.

Additional Compostable Materials

Besides the traditional greens and browns, other household and garden materials can enrich your compost:

Believe it or not, - **Hair and fur**: Human and pet hair are high in nitrogen and decompose well over time.

- **Natural fibers**: Cotton and wool rags, rope, and string, as long as they are not dyed or coated with synthetic materials.

- **Nut shells**: Particularly walnut, pecan, and almond shells, which decompose slowly but add grit and texture.

It's worth noting that - **Heirlooms and stale bread**: Stale bread can be composted, provided it is broken into small pieces.

Avoid moldy bread as it can introduce unwelcome fungi.

What to Avoid in Your Compost

While many organic materials are compostable, certain items must be avoided to ensure an effective and healthy compost pile:

It's worth noting that - **Meat, fish, and dairy**: These can attract pests and produce foul odors as they decompose.

- **Fats and oils**: These inhibit the breakdown process by coating other materials and impeding aerobic decomposition.

- **Disease-prone plants**: Avoid adding plants affected by diseases, such as blight or rust, to prevent spreading to your garden.

In real life, - **Invasive weeds**: Some weeds, like bindweed or ivy, can survive the composting process and could invade your garden when applied as compost.

- **Chemically-treated wood**: This includes sawdust from pressure-treated or painted wood due to toxic substances.

Ratio and Layering Tips

Achieving the right balance of materials is key to successful composting. Ideally, your compost pile should consist of approximately three parts browns to one part greens.

Layering these materials akin to a lasagna will help maintain this balance. Start with a base layer of coarse browns, such as twigs or straw, to provide aeration. Follow with a layer of greens, then alternate layers ensuring browns are the final layer to deter pests.

Troubleshooting Common Compost Problems

- **Smelly Compost**: This usually results from excess greens or poor aeration. Add more browns and mix thoroughly to incorporate air.

- **Slow Decomposition**: If your compost seems inactive, it could be due to a lack of greens. Introduce nitrogen-rich materials and ensure adequate moisture.

Conclusion

Composting is an invaluable skill that transforms waste into a garden's best ally. By knowing which materials to compost, and in what proportions, you foster an ecosystem that effectively recycles nutrients.

With this knowledge in hand, you're well-equipped to build a thriving compost pile that contributes both to reducing household waste and enriching your garden's soil. As you continue composting, you'll become more attuned to which materials align best with your needs, ultimately contributing to a more sustainable lifestyle.

Chapter 8: What Not to Compost: Avoiding Common Contaminants

Composting is a fantastic way to convert kitchen scraps and yard waste into rich, fertile soil that can enhance your garden's health and productivity. The practice not only aids in reducing landfill waste but also recycles nutrients to create this natural fertilizer.

However, while almost all organic materials can decompose under the right conditions, not all materials are suitable for composting. Ensuring that you only add appropriate materials to your compost pile is crucial to maintaining a balanced, healthy system. This chapter covers what not to compost, explaining why certain materials can do more harm than good when tossed into your compost bin.

1. Meat, Bones, and Fish Scraps

I've seen this happen before: Although they decompose naturally, meat, bones, and fish scraps should be avoided in your compost pile. The primary concern with these materials is the attraction of pests such as rodents, raccoons, and flies, which can become serious nuisances.

Moreover, these materials decompose anaerobically and can create bad odors, interrupting aerobic bacteria's activity in your compost. The oils and fats present in these materials can also delay the decomposition process.

2. Dairy Products

Similar to meat and fish, dairy products like milk, cheese, and yogurt can create odors and attract pests.

Additionally, their high moisture content can upset the balance of your compost pile. Adding dairy can lead to a slimy texture, disrupting the aeration and yellowing the balance needed for effective decomposition.

3. Fats, Grease, and Oils

Fats, grease, and oils from cooking should be kept out of your compost.

These substances can coat other compost materials, preventing sufficient air circulation and slowing down the composting process. They also contribute to foul odors and serve as attractants to pests, much like meat and dairy products.

4. Diseased Plants

Interestingly enough, While adding garden waste to your compost is generally beneficial, you should avoid plants that show signs of disease.

Diseases, fungi, and viruses can survive the composting process, especially if it doesn't reach consistent high temperatures necessary for sterilization. Introducing these pathogens into your garden with your compost can result in the spread of diseases to healthy plants.

5. Weeds Gone to Seed

Composting is effective in killing most weed seeds, but this often requires consistent high temperatures that many home compost piles don't achieve.

If you add weeds that have gone to seed, you run the risk of spreading these weeds when you use the compost. To be safe, avoid adding mature weeds to your pile.

6. Treated Wood Products

Though wood typically makes an excellent addition to compost as a source of carbon, any wood that has been treated or painted should be excluded.

Treated wood and painted surfaces can contain harmful chemicals, including heavy metals and toxins like arsenic. Composting these materials can contaminate your compost and pose risks to your garden's health and safety.

7. Pet Waste

It's worth noting that Feces from dogs, cats, and other carnivorous pets contains pathogens and parasites that can transmute diseases such as Toxoplasmosis and various intestinal worms.

Unlike herbivore manure, which can be composted, pet feces require specific conditions for safe decomposition, making them unsuitable for typical backyard composting.

8. Glossy or Coated Paper

While paper is generally compostable, it's essential to be selective about the type you use.

Avoid glossy paper, such as magazines or brochures, which are often treated with chemical dyes and coatings that may take longer to break down and potentially introduce toxins. Stick with plain, non-colored, non-glossy paper for your compost.

9. Large Amounts of Citrus Peels and Onions

Citrus peels and onions have an acidic nature which can slow down microbial activity in your compost, prolonging decomposition time.

Though small amounts are generally okay, be wary of adding them in large quantities, which could create an unbalanced compost environment. Additionally, their strong odors might also deter beneficial insects and microorganisms.

10. Coal Ash

Unlike wood ash, which can be a useful compost additive, coal ash should be avoided.

Coal ash contains high levels of sulfur and iron, which can make the soil too acidic and potentially introduce heavy metals into your compost, consequently affecting soil health and plant growth.

11. Synthetic and Biodegradable Plastics

There is a growing trend towards using biodegradable plastics, but many of these are not suitable for home composting.

Such materials often require industrial composting facilities to break down completely. Including synthetic or even some 'compostable' plastics in your pile can lead to a littered compost, leaving behind residues that do not decompose adequately in most backyard systems.

12. Minimal Non-Organic Material

As a rule of thumb, non-organic materials should be entirely avoided.

Plastics, metals, and synthetic fabrics do not break down in compost. Even if some items are small or seem negligible, they can persist in your compost long after the organic material has decayed, potentially contaminating the compost with physical debris.

In conclusion, knowing what not to compost is just as vital as knowing what should go into your pile. By avoiding these common contaminants, you help ensure that your composting endeavors result in a nutrient-rich product that will nourish your plants rather than harm them.

A little vigilance today will save you headaches in the future, preserving the benefits of a greener, more sustainable lifestyle. Keep your compost pile both productive and pest-free by following these guidelines, and you'll be well on your way to creating a thriving garden paradise.

Chapter 9: The Perfect Balance: Nitrogen vs. Carbon

Understanding the balance between nitrogen and carbon is crucial for any successful composting endeavor. It forms the backbone of the process, determining the speed and quality of compost formation.

Delving into the world of composting reveals that decomposing organic matter is not just about tossing kitchen scraps into a pile; it involves a delicate dance of ratios and natural processes. Achieving the perfect balance of these key elements can make the difference between a thriving compost heap and a sluggish, ineffective one.

The Role of Nitrogen and Carbon

Many people don't realize this, but At its core, composting is the natural decomposition of organic matter, and it requires two primary components: nitrogen-rich materials, often referred to as "greens," and carbon-rich materials, known as "browns.

" These two elements serve unique purposes within the compost pile.

Nitrogen is vital because it provides proteins and amino acids, which are essential for the growth of micro-organisms responsible for decomposing organic matter.

Greens, such as kitchen scraps, grass clippings, and manure, are high in nitrogen content. These materials stimulate the microorganisms, acting as an energy source, speeding up the decomposition process.

On the other hand, **carbon** is crucial for providing energy for these microorganisms.

Carbon-rich materials include dried leaves, straw, and wood chips. Browns are fundamental in creating the bulk of the compost, providing structure to the pile, ensuring aeration, and preventing compaction, which can lead to anaerobic conditions that slow decomposition and produce odors.

The Crucial Carbon to Nitrogen Ratio

A successful compost pile thrives on the right carbon-to-nitrogen (C:N) ratio.

The optimal ratio for composting is generally around 30:1 (carbon to nitrogen). Attaining this balance ensures that microorganisms have enough carbon for energy and nitrogen for building proteins, creating an environment where decomposition occurs efficiently.

Practical Steps to Achieve the Perfect Ratio

I've seen this happen before: To achieve an optimal C:N ratio, begin by classifying your available organic materials into greens and browns.

Remember that appearance can sometimes be deceiving. For example, coffee grounds are classified as a green due to their high nitrogen content, even though they are brown in color.

Once you have separated your materials, consider their C:N ratios.

Grass clippings, for instance, have a ratio of about 15:1, whereas fallen leaves might have a ratio of 60:1. When you mix these materials, aim for a blend that results in an overall ratio close to the 30:1 benchmark.

Layering is an effective technique for achieving balance. Start with a layer of coarse carbon materials like branches or straw to encourage aeration.

Follow this with a layer of greens such as kitchen scraps or grass clippings, then cover with a layer of shredded leaves or small branches. Repeating this layering process as you add more waste ensures a balanced and aerated pile.

Adjusting the Ratio

There may be times when you need to adjust the composition of your pile.

A compost heap that emits a foul odor or is soggy often indicates an excess of nitrogen, requiring more carbon materials to balance it out. Conversely, if the decomposition process seems slow and the pile is too dry, it suggests a lack of nitrogen, necessitating the addition of more greens.

Additionally, if the texture of the pile appears matted or compacted, incorporating coarse brown materials can help enhance aeration and prevent the formation of an anaerobic environment.

Troubleshooting and Tips

1. **Observation**: Regularly monitor the pile for signs of imbalance.

Smell, texture, and heat are all indicators of how well your compost pile is doing. A healthy compost pile will have a pleasant, earthy smell and should feel warm to the touch.

2.

Aeration: Turning the compost pile every few weeks introduces oxygen, which is imperative for sustaining aerobic microorganisms. This helps maintain the optimal C:N ratio by evenly distributing the materials and enhances the breakdown process.

Many people don't realize this, but 3. **Moisture Balance**: The moisture content of your compost is just as crucial as the C:N ratio.

The pile should be as moist as a wrung-out sponge. If it's too dry, microorganisms cannot function effectively, and if it's too wet, the pile runs the risk of becoming anaerobic and smelly.

Benefits of a Balanced Compost

A well-balanced compost pile results in numerous benefits.

It produces rich, dark compost teeming with nutrients and beneficial organisms that can enrich garden soil, improving its texture, water retention, and ability to support plant growth. By maintaining the perfect balance of nitrogen and carbon, gardeners can ensure a quick and efficient composting process, turning waste into a valuable resource that nourishes their plants naturally.

Conclusion

In the journey to create the perfect compost, understanding and implementing the correct nitrogen-to-carbon balance is paramount. By utilizing the techniques of careful observation, correct layering, and periodic adjustments, composters can harness the power of natural decomposition effectively.

The resulting compost not only reduces waste but also rejuvenates gardens, promoting a sustainable cycle of growth and renewal. As you hone your composting skills, remember that this art is as much about learning and adapting as it is about precise ratios—a perfect embodiment of nature's innate balance.

Chapter 10: Managing Moisture: The Key to Decomposition

In the intricate dance of turning kitchen scraps and yard waste into rich, garden gold, moisture plays a fundamental role as both a facilitator and a regulator of the composting process.

Much like a sponge that thrives best neither parched nor waterlogged, a compost pile demands an optimal moisture balance to efficiently break down organic matter.

Understanding the Role of Moisture

Compost is essentially a teeming ecosystem of microorganisms — bacteria, fungi, and insects — that work in harmony to decompose organic material. These microscopic decomposers are akin to other living organisms in their essential need for water.

They rely on moisture to thrive and function effectively. Water acts as a solvent, enabling bacteria and fungi to break down organic matter at a molecular level. It also facilitates the metabolic processes within these microbes, ultimately speeding up the composting cycle.

The Sweet Spot: Ideal Moisture Levels

The target moisture level in a compost pile generally mirrors that of a wrung-out sponge — sufficiently moist but not dripping wet. This typically translates to a moisture content of approximately 40-60%.

If the compost becomes too dry, the microbial activity can slow to a near halt, prolonging the decomposition process. Conversely, if the pile is too wet, the excess water can create anaerobic conditions, which might lead to unpleasant odors and the production of harmful compounds.

Measuring Moisture Content

While specific moisture meters are available to ascertain the exact water content within your compost, these tools aren't necessary for every composter. A simple squeeze test can be just as effective: grab a handful of compost from the middle layer of your pile.

Squeeze it firmly in your hand. If just a few drops of water emerge, your compost likely maintains an ideal moisture level. If it feels like wringing out a soaked sponge, it's too wet, while dry and crumbly means it's too dry.

Adjusting Moisture Levels

In real life, Balancing moisture begins with materials.

Browns, like dried leaves, straw, and paper, tend to be dry, while greens, such as vegetable scraps and grass clippings, are moisture-rich. To adjust, consider the following steps:.

- **To Increase Moisture:** Incorporate more greens into your mix.

Fresh grass clippings, fruit peels, and other wet kitchen scraps will naturally add water to the pile. Sprinkling water with a hose or watering can, especially while turning the pile, can also help distribute moisture evenly.

- **To Decrease Moisture:** Add more browns.

Dry leaves, newspaper, and cardboard will absorb excess water. Additionally, ensuring your pile is aerated will help evaporate some of the excess moisture.

Aeration: Complementing Moisture Management

Adequate aeration is a vital counterpart to managing moisture.

A well-aerated pile supports aerobic microorganisms — those that require oxygen — and prevents the growth of anaerobic microbes, which are often the culprits behind foul odors. Regular turning of the compost helps achieve this by introducing air into the pile, thereby supporting microbial activity and preventing anaerobic conditions that excessive moisture might otherwise foster.

Seasonal Considerations

Climate and seasonality significantly impact moisture management. During rainy seasons, protect your compost by covering it with a tarp or placing it under a shelter.

This prevents waterlogging and maintains the structural integrity of the pile. In contrast, hot and dry climates may require more frequent watering or additional incorporation of wet materials to keep the compost sufficiently moist.

The Risks of Improper Moisture Management

Leaving a compost pile to its own devices without proper moisture control can lead to several issues.

Overly dry piles often invite pests, such as rodents, seeking a dry home. Wet and compacted compost, on the other hand, may become a breeding ground for anaerobic bacteria, leading to a swamp-like, smelly heap.

Troubleshooting Common Issues

- **Smelly Compost:** Usually indicative of too much water or lack of oxygen. Add dry carbon materials and turn the pile to aerate.

- **Slow Decomposition:** Often a sign of insufficient moisture. Test the pile's moisture content and add water or greens as needed.

Establishing a Maintenance Routine

Develop a routine to monitor and adjust moisture levels.

Regular checks — weekly, for most hobbyists — can preemptively address potential issues. Consider keeping a log or schedule as part of your composting practice to track any changes or adjustments made over time, which can be crucial for refining your technique season after season.

Conclusion: The Balancing Act

Tending to a compost pile is an act of balance and observation, where moisture holds the transformative potential to accelerate decomposition or hinder it entirely. Mastering moisture management is not just about addressing immediate challenges but embracing an ongoing learning process that benefits both your compost and garden.

By understanding the pivotal role of water and diligently maintaining the optimal levels, you pave the way to rich, fertile soil and sustainable gardening practices. Embrace the nuances, trust in the process, and watch as your waste turns into an elemental resource, providing endless nourishment for your garden.

Chapter 11: Turn and Aerate: Why Oxygen is Crucial

As you journey deeper into the world of composting, you've likely come across the term "aeration" multiple times. Its frequent mention underscores its importance in the composting process.

Turning and aerating your compost pile is not just a procedural step; it's a crucial action that transforms a mere heap of organic waste into rich, life-giving humus. This chapter seeks to elucidate the role of oxygen in composting and provide practical guidance on how best to introduce and manage airflow within your compost.

The Science Behind Oxygen in Compost

In real life, To fully appreciate the necessity of oxygen, we need to understand the biological engine that drives composting: the microbes.

These tiny, diligent workers are primarily aerobic, meaning they need oxygen to metabolize organic material. In the presence of oxygen, these microbes efficiently break down your kitchen scraps, lawn clippings, and other organic matter, producing heat, carbon dioxide, and water vapor as byproducts, eventually converting the mass into compost.

When oxygen is plentiful, composting is a swift and odorless process. However, depriving these microorganisms of oxygen forces anaerobic bacteria to take over.

Anaerobic decomposition is much slower and, more importantly, it produces unpleasant odors, such as hydrogen sulfide, and methane—a potent greenhouse gas. To avoid this, maintaining aerobic conditions is essential.

Observing and Understanding Your Compost Pile

At first glance, Before diving into how to aerate, it's important to understand when your pile needs it. There are several telltale signs that your compost might be starved of oxygen.

If you notice a rotten smell, that indicates anaerobic activity. A pile that seems cool to the touch may also lack oxygen, as heat generation is a sign of vigorous aerobic activity. Additionally, if your compost pile seems overly wet and compacted, airflow is likely insufficient—oxygen has a hard time penetrating dense masses.

Tools and Techniques for Aeration

Now that we understand why oxygen is important, let's explore how to ensure your compost pile is adequately aerated.

Here are some practical tools and techniques:.

Many people don't realize this, but 1. **Turning with a Pitchfork or Shovel**: One of the simplest and most effective ways to aerate your compost is to turn it regularly.

Use a pitchfork or shovel to manually mix the layers of your pile. This action introduces air and redistributes materials, promoting even decomposition. Aim to turn your pile every week or every couple of weeks, depending on your composting system and material mix.

2. **Compost Aerators**: These are specialized tools designed to introduce air into your compost effortlessly.

They often feature spirals or wings that penetrate the pile and, as you pull them back, create air channels. They are especially useful for those who may find turning a large pile with a pitchfork physically demanding.

3.

Perforated Pipes: A more passive method involves inserting vertical or horizontal pipes within the pile. These pipes, perforated with holes, allow air to circulate through the core of the compost mass, ensuring oxygen reaches even the innermost parts.

At first glance, 4.

Layering: Building your pile in a manner that naturally allows air circulation can reduce the frequency of turning. Alternate between green materials (rich in nitrogen) and brown materials (rich in carbon), as well as bulky material like sticks or straw that can help create air pockets.

5.

Choosing the Right Bin: Consider the design of your compost bin. A bin with built-in ventilation features, such as side slats or a lid design that promotes airflow, can help maintain aerobic conditions with less effort.

Additional Tips for Effective Aeration

- **Monitor Moisture Levels**: Compost should be as moist as a wrung-out sponge.

Too much water can fill air spaces and exclude oxygen. If your pile feels soggy, add more brown material and turn it to introduce air.

- **Optimize Pile Size**: An ideal compost pile is large enough to retain heat but not so large that oxygen cannot penetrate its interior.

A size of 3x3x3 feet is often recommended for maintaining a balance of heat retention and aeration capacity.

- **Break Down Materials**: Shred larger materials before adding them to your pile.

Smaller pieces decompose faster and allow for better air flow compared to large, compacted masses of organic matter.

The Benefits of Proper Aeration

Effectively aerating your compost has a range of benefits that extend beyond simply managing odors and warding off anaerobic bacteria.

When your pile has adequate oxygen, composting occurs more rapidly, meaning you'll have nutrient-rich compost ready to nourish your garden sooner. Additionally, heat-loving (thermophilic) bacteria thrive, reaching temperatures high enough to kill weed seeds and pathogens, enhancing the quality of your compost.

Troubleshooting Common Issues

I've seen this happen before: Despite our best efforts, issues may still arise.

If your compost pile smells or seems inactive despite regular turning, consider the following:.

- **Check the Balance**: The nitrogen to carbon ratio might be off.

Excess nitrogen can cause odor, while too much carbon slows the process. Aim for a balance where nitrogenous greens make up about one-fourth to one-third of the material in your compost pile.

- **Examine Weather Conditions**: Extended periods of rain can saturate your pile, while very dry weather might strip it of necessary moisture.

Adjust accordingly by covering your pile to shield it from excessive rain or watering it during dry periods.

Remember, composting is as much an art as it is a science. It demands observation, experimentation, and patience.

Each pile is a little different based on a myriad of factors including the materials used and your local climate conditions. By ensuring proper aeration, you are taking one of the most important steps towards successful composting, ultimately yielding a garden that is vibrant, healthy, and sustainable.

Chapter 12: Troubleshooting Common Compost Problems

While composting is a straightforward process, it is not without its occasional hiccups.

Understanding how to identify and remedy common composting problems is key to maintaining a healthy bin and producing quality compost. This chapter will guide you through diagnosing and solving typical issues you might encounter.

Odor Problems

Problem: The Compost Smells Bad

An unpleasant odor is one of the most common complaints among composters.

A foul smell generally indicates an oxygen deficiency, excessive moisture, or a lack of balance between green and brown materials.

Solution:

1.

Aerate the Compost: Regularly turning your pile ensures that air circulates throughout, preventing anaerobic conditions that cause odor. If your compost smells like rotten eggs, it's time to break out the pitchfork and give it a good turn.

2. **Adjust Moisture Levels:** A soggy pile can quickly become smelly.

Ensure your compost is as moist as a wrung-out sponge. If it's too wet, add dry brown materials like shredded newspaper or sawdust to absorb excess water.

To be honest, 3.

Balance Green and Brown Materials: An overabundance of nitrogen-rich greens (like food scraps and fresh grass clippings) without enough carbon-rich browns (like dry leaves and straw) can lead to odors. Strive for a 2:1 ratio of browns to greens for optimal decomposition.

Temperature Issues

Problem: The Compost Pile is Not Heating Up

A compost pile that fails to heat up may not be decomposing effectively.

Since heat is a byproduct of active microbial activity, a cool pile indicates sluggish microbial action.

Solution:

In real life, 1.

Check the Pile's Size: A very small pile may struggle to heat up. Aim for a minimum of 3x3x3 feet to create a critical mass that retains heat well.

2.

Add More Greens: If your pile is cool and dry, mix in more nitrogen-rich greens. These materials provide the energy microbes need to thrive and raise the pile's temperature.

3. **Insulate the Pile:** In colder weather, covering your compost pile with a tarp can help retain heat and protect it from cooling breezes.

Moisture Problems

Problem: The Compost is Too Dry

Dry compost can stall decomposition and lead to a loss of microbial life.

Solution:

1.

Moisten the Pile: Gradually add water to your compost, turning it as you go to ensure even distribution. The goal is to achieve a uniform moisture content similar to a damp sponge.

As strange as it sounds, 2. **Add Greens or Wet Browns:** Incorporate water-rich materials like fruit peels or tea bags to boost moisture levels.

Problem: The Compost is Too Wet

Excess moisture can suffocate bacteria and lead to an anaerobic, smelly mess.

Solution:

1. **Turn the Pile Broadly:** This will help to release trapped moisture and introduce fresh air.

At first glance, 2.

Mix in Dry Materials: Include a healthy amount of shredded leaves, straw, or cardboard to absorb the excess water and regain balance.

Pest Problems

Problem: The Compost Attracts Rodents or Other Pests

Believe it or not, Presence of pests can be discouraging and disruptive to composting efforts.

Solution:

1. **Avoid Animal Products:** Do not add meats, fats, or dairy products to your compost. These are major pest attractants.

2.

Utilize a Compost Bin with a Lid: A securely closed bin can prevent pests from entering. Opt for bins with small holes that allow for air circulation but block larger creatures.

3. **Bury Food Scraps:** Always cover new kitchen scraps with a layer of brown materials.

Decomposition Problems

Problem: The Compost is Not Breaking Down

Inadequate decomposition can result from multiple factors, such as poor mixture, incorrect moisture level, or insufficient aeration.

Solution:

1.

Shred Large Pieces: Break down large chunks of organic matter to speed up decomposition. Smaller pieces have a greater surface area for microbes to work on.

Believe it or not, 2. **Rebalance Your Ingredients:** Ensure you have a good mix of greens and browns.

3. **Turn the Pile Frequently:** Turning helps combine materials, aerate the pile, and distribute moisture evenly.

Miscellaneous Problems

Problem: Invasion of Weeds

Weeds in a compost pile usually result from adding their seeds unchecked into the mix.

Solution:

1.

Pre-compost in a Hot Pile: Make sure your compost pile gets hot enough to kill weed seeds (130-150°F for several days). Frequent turning helps maintain these high temperatures.

2.

Avoid Adding Seed-Starting Materials: Be careful about adding plants that have gone to seed. Compost them only if you are certain your pile will reach the necessary temperatures.

Conclusion

A successful composting experience is often a result of close observation and timely adjustments. Although it may feel daunting initially, as you solve these minor issues, you'll grow more confident and adept at managing your compost.

Remember, each problem is an opportunity for learning. With patience and practice, you'll soon be producing rich, aromatic compost to nourish your garden organically.

Chapter 13: Speeding Up the Composting Process Naturally

Composting is a process that Nature has perfected over millions of years, turning waste into a resource through the slow and steady work of microorganisms.

However, in our fast-paced world, many gardeners aspire to expedite this process without compromising the quality of the compost. This chapter will guide you through natural methods to accelerate composting so you can enjoy nutrient-rich compost sooner.

Understanding the Basics

At first glance, To hasten composting, it's vital to grasp the basics of what turns food scraps and garden waste into compost. The process relies on microorganisms like bacteria and fungi, which break down organic material. These organisms require four key elements: carbon, nitrogen, oxygen, and moisture.

Carbon-rich materials include dried leaves and woody stems—often referred to as "browns. " Nitrogen is provided by green materials like grass clippings and kitchen scraps. Oxygen is crucial for them to breathe and thrive, while moisture is necessary for decomposition but should be balanced to prevent the pile from becoming waterlogged.

Particle Size Matters

One of the simplest ways to speed up the composting process is by reducing the size of the materials being composted. The larger the surface area exposed to decomposers, the faster they can work.

Chopping or shredding items like yard waste and kitchen scraps increases the surface area and offers more access points for the microorganisms. A wood chipper is perfect for garden debris, while a kitchen blender can finely chop soft kitchen scraps. By making the initial material as accessible as possible, you're setting up a banquet feast for decomposers rather than a long, drawn-out dinner.

Balancing the Green and Brown

Ensuring the proper ratio between nitrogen-rich greens and carbon-rich browns is crucial—often cited as a ratio of 1:30 by weight. Too much nitrogen makes the pile smelly and anaerobic, while too much carbon slows down the process.

If your compost pile seems to be stagnating, adjust the balance by adding more greens like coffee grounds or grass clippings, or if it's too wet and smelly, add browns like straw or shredded newspaper. Tracking your proportions in a compost diary can offer valuable insights over time, helping you fine-tune the perfect mix for rapid decomposition.

Layering for Success

Layering compost materials in a strategic way can have a significant impact on the speed of decomposition. The idea is to alternate layers of green and brown materials, starting and finishing with brown.

Consider including a layer of fresh garden soil midway as an inoculant to introduce even more beneficial microorganisms to feast upon the organic matter. This stratification encourages movement, ensuring moisture and aeration are evenly distributed.

Maintaining Optimal Conditions

Temperature is a critical factor in the speed of composting. A pile's temperature should ideally be between 130°F and 160°F.

This temperature range promotes rapid decomposition and kills off many weed seeds and pathogens. You can increase the pile temperature by making your compost pile large enough—about three feet by three feet by three feet is the minimum ideal size. Ensure the pile receives enough sun to stay warm but be careful not to let it dry out, maintaining it at the moisture level of a damp sponge.

As strange as it sounds, If the temperature drops significantly, especially in winter or rainy seasons, a compost thermometer can help you monitor and take corrective action.

If it's too cool, turning the pile or adding more greens can stimulate microbial activity. In colder regions, thermal blankets can help preserve heat.

Turning and Aeration

Turning the compost pile regularly introduces oxygen, which accelerates the decomposition process. Aeration is vital as our active decomposers require oxygen.

For a faster turnover, implement a consistent turning schedule—every couple of weeks is often recommended. For a gardener in a hurry, turning weekly or even bi-weekly can yield finished compost much sooner as it invigorates aerobic bacteria.

At first glance, The method of turning doesn't have to be labor-intensive.

Tools like a pitchfork or compost aeration tool can make the task easier. In smaller bins, simply lifting the contents from the bottom and turning them over can also suffice.

Keeping Moisture Levels Just Right

Moisture is essential for composting since microorganisms need a moist environment to thrive. However, too much moisture leads to a lack of air and can turn your pile anaerobic.

Conversely, if the pile is too dry, microbial activity slows down. To check the moisture level, take a handful of the composting material and squeeze it. It should feel moist to touch, with only a few drops of water being expelled.

Let's not forget, In dry periods, watering cans or a hose with a fine spray can remedy dryness.

Building a roof or covering with a tarp can stave off excess rainwater. The idea here is balance—with practice, you will develop an intuitive sense of what your compost pile needs.

Using Activated Starters and Additives

Another effective strategy involves adding compost boosters such as manures, commercial activators, or effective microorganisms (EM).

These amendments can replenish microorganism populations and accelerate the breakdown processes. While these should not replace sound composting practices, they can be a useful supplement when trying to speed up the process.

I've seen this happen before: When opting for commercial starters, ensure that they're as natural as possible, avoiding synthetic chemicals which may imbalance the soil's natural ecosystem when the compost is used.

Conclusion

By incorporating these strategic yet simple recommendations, you'll find your compost is coming along faster than ever before—ready to enrich your garden and nourish your plants. Remember, while a quick turnover is desirable, achieving nutrient-rich, balanced compost should always be the paramount goal.

Like the deep-rooted wisdom of Nature's cycles, the art of speeding your compost process naturally lies in harmony, patience, and an observant, responsive touch. Welcome to the accelerated world of composting, where gardens flourish, wastes diminish, and sustainable practices thrive.

Chapter 14: Seasonal Composting Tips and Tricks

Composting is a dynamic process that requires different strategies throughout the year to optimize the breakdown of organic materials. Each season presents unique challenges and opportunities for composting enthusiasts.

By adapting your composting practices to suit these seasonal variations, you can enhance the efficiency of your compost pile and produce nutrient-rich humus for your garden. This chapter offers a collection of tips and tricks to help you make the most of your composting efforts, no matter the season.

Spring: Kick-Start Your Compost

At first glance, As the earth awakens from winter's grasp, spring is the perfect time to revitalize your composting routine.

Warmer temperatures and increased moisture create ideal conditions for microbial activity. Here's how to take advantage of spring's renewal:.

1. **Turn Over a New Leaf**: Begin by turning your compost pile thoroughly.

This aerates the material, breaks up compacted layers, and introduces oxygen, which is critical for aerobic decomposition. Use a pitchfork to mix the materials and encourage faster breakdown.

2. **Balance the Components**: Spring often brings an abundance of green waste, such as grass clippings and fresh plant material.

Balance this nitrogen-rich material with carbon-rich brown materials like dried leaves, straw, or shredded newspaper to maintain an optimal carbon-to-nitrogen ratio of about 30:1. This balance prevents the pile from becoming too moist or too dry.

3. **Moisture Maintenance**: Spring showers can saturate your compost pile.

Ensure your compost has a sponge-like consistency by covering it during heavy rainfalls. An old tarp or plastic sheet can protect the pile and maintain the ideal moisture level, enabling microbes to thrive.

4. **Add Activators**: To jump-start composting, consider adding a compost activator.

This can be commercial activator products or natural options like manure, coffee grounds, or alfalfa meal. These add nitrogen and beneficial microbes, speeding up the decomposition process.

Summer: Managing the Heat

Interestingly enough, Summer heat catalyzes microbial activity, accelerating the composting process.

However, it also poses challenges such as excessive heat and drying out. Here's how to manage your compost through these peak months:.

1. **Hydration Vigilance**: Hot weather can dry out your compost, slowing down decomposition.

Regularly check moisture levels, ensuring the pile is kept as damp as a wrung-out sponge. Water the pile when needed or incorporate water-rich food scraps to maintain adequate hydration.

2. **Shade Solutions**: If possible, situate your compost bin or pile in a shaded area to prevent overheating.

Excessive temperatures can kill beneficial microbes. If relocating isn't feasible, cover the pile during the hottest part of the day to shield it from direct sunlight.

Many people don't realize this, but 3.

Regular Aeration: High temperatures accelerate decomposition, causing the pile to compact more quickly. Turn the pile every few weeks to maintain adequate aeration, which promotes microbial activity and helps distribute heat evenly.

4. **Pest Prevention**: Summer can also mean an increase in pest activity.

Keep food scraps buried under brown material, such as leaves or straw, to minimize odors and deter rodents or insects. Consider using pest-proof compost bins for added security.

Autumn: Harvesting and Layering

Many people don't realize this, but Autumn is a bountiful season for composting, as gardens offer a wealth of organic material.

Leaves, spent plants, and garden trimmings abound. Here's how to handle this autumnal abundance:.

1. **Leaf Management**: Fallen leaves are a fantastic carbon source.

Shred leaves before adding them to your compost to speed up their decomposition. Use a lawnmower with a bag attachment or a leaf shredder for this task.

2. **Layering Techniques**: The variety of materials available in autumn is perfect for building a layered compost pile.

Alternate layers of green materials, like grass clippings and vegetable scraps, with brown layers, like shredded leaves and straw. This stratification encourages even break down.

It's worth noting that 3.

Final Harvest: Before the first frost, harvest the finished compost from the bottom of your bin or pile. Use this nutrient-rich soil amendment to prepare your garden beds for winter, adding a layer of free organic matter that will enhance soil fertility come spring.

4.

Insulate for Winter: As temperatures begin to drop, consider insulating your pile to extend the composting season. Use straw bales, tarps, or thick layers of leaves as insulating material around the pile to retain warmth and keep microbes active longer.

Winter: Maintain the Momentum

In real life, Composting slows down in winter but doesn't' necessarily stop.

By adjusting your approach, you can keep decomposition ticking over through the colder months:.

1.

Size Matters: A larger compost pile retains heat better, so consider combining smaller piles into a single, larger mass. This thermal mass keeps the core of the pile warm, sustaining microbial activity.

2.

Enhanced Insulation: Further insulate your pile using more layers of straw or leaves, or even construct a windbreak if your compost is exposed to the cold. Keeping the pile insulated prolongs the active composting period.

Believe it or not, 3.

Indoor Options: If outdoor composting is impractical, consider indoor composting solutions like worm bins or electric composters. These methods allow for year-round composting regardless of outside conditions.

4.

Additions in Winter: Continue adding food scraps to the pile, but it's advisable to chop or shred these inputs as much as possible to facilitate quicker breakdown. Be prepared for a slower decomposition process and manage your expectations accordingly.

By applying these seasonal composting tips and tricks, you not only improve the efficiency and speed of your composting but also ensure that you have a continuous supply of rich, organic compost ready to nourish your garden year-round.

Adapting to seasonal changes not only enhances the quality of the compost produced but also strengthens your role in recycling waste into valuable organic matter. As you integrate these strategies into your routine, you'll find that composting becomes not just an occasional activity, but a seamless part of your seasonal gardening cycle.

Chapter 15: Harvesting and Using Finished Compost

Composting is a journey that transforms garden and kitchen waste into a rich, nutrient-dense amendment that enhances soil health and supports vibrant plant growth.

As gardeners and environmental stewards, we eagerly anticipate the moment when our compost is ready to harvest and enrich our gardens. In this chapter, we'll explore how to determine if your compost is finished, the best methods for harvesting it, and the various ways you can apply it to benefit your garden.

Determining When Compost is Ready

Before harvesting your compost, it's essential to confirm that it's fully decomposed. Mature compost, also known as humus, has a dark, crumbly texture and resembles rich soil.

It should have an earthy aroma, free of any foul or rotten smells, which indicates the decomposition process is complete. If you notice recognizable food scraps or coarse organic matter, your compost may need more time to break down further. Depending on the composting methods used—hot or cold—the timeline for completion can vary, particularly influenced by the pile's maintenance and environmental conditions.

A quick test to determine readiness is the "bag test".

Place a small amount of compost in a sealed bag for a week and then open it. If it smells like dirt, it's ready; if it has an ammonia or sour smell, your compost needs more time.

Harvesting Compost

Interestingly enough, Once you're confident that your compost is ready, it's time to harvest.

Harvesting methods will depend on the type of composting system you've employed.

1. **Traditional Compost Pile/Bin**: If you're using a traditional pile or bin, begin at one edge and work your way down.

A garden fork or spade is an excellent tool for this task. Separate any unfinished material you might encounter and return it to the bin for further decomposition.

2. **Tumblers**: For compost tumblers, simply rotate the tumbler above a wheelbarrow or tarp.

Mature compost will easily pour out. As with piles, be on the lookout for partially decomposed materials to recycle into a new batch.

In real life, 3. **Vermicomposting**: Harvesting from a worm bin requires a bit more finesse.

An effective method is to feed one side of the bin for a few weeks, which will attract worms to that area. You can then remove the compost from the untouched side, ensuring fewer worms are disturbed. Alternatively, light exposure will drive worms deeper, allowing the surface compost to be collected easily.

4.

Compost Bins with Removable Sections: Some bins have a door or detachable section at the bottom for easy harvesting. Simply open and remove the compost without disturbing the upper layers still undergoing decomposition.

Sieving Compost

Let's not forget, Once harvested, it's beneficial to sieve your compost to create a finer, more uniform product. Use a compost sieve or a mesh screen with holes about a half-inch in diameter.

This process removes larger undecomposed materials, which can be returned to your compost pile. Fine, sieved compost is particularly useful for starting seeds or incorporating into potting mixes.

Storing Finished Compost

If you're not ready to use your compost immediately, you can store it for later use. Keep it in a dry, covered area to prevent nutrient leaching.

Moisture and heat can instigate further decomposition, potentially transforming your compost into a dense, anaerobic mess. Ensure your storage solution allows for some airflow, helping to maintain its quality until you're ready to use it.

Using Compost in Your Garden

Applying compost to your garden can revolutionize your soil's health and productivity. Here's how you can best utilize it:

1.

Soil Amendment: Mix compost into garden beds to improve soil structure, aeration, and drainage. Aim for 2-3 inches of compost on the surface, then integrate it into the top 6-8 inches of soil.

Let's not forget, 2.

Top Dressing: Spread a 1-inch layer of compost around existing plants as a mulch. This not only supplies plants with nutrients as the compost breaks down but also helps retain soil moisture and suppress weeds.

3.

In Container Gardens: Add compost to your potting mix. A ratio of one part compost to three parts soil works well, improving moisture retention and nutrient content for potted plants.

4. **Compost Tea**: Brewed from finished compost, compost tea provides a liquid fertilizer rich in soluble nutrients and beneficial microbes.

Steep a shovelful of compost in a bucket of water, aerating for 24-48 hours. Strain and use the liquid to water plants at the roots or as a foliar spray.

Interestingly enough, 5. **Lawn Care**: Compost is excellent for lawns, serving as a natural fertilizer.

Apply a thin layer, about ¼ inch, and work it gently into the grass. This promotes healthier turf, supports beneficial soil organisms, and reduces the need for chemical fertilizers.

Considerations for Specific Plants

While most garden plants thrive with compost, some have specific needs.

Acid-loving plants, like blueberries, may not require as much compost, so adjust the amount you add. Furthermore, when growing root vegetables, ensure compost is well-integrated into the soil to avoid deformities.

Conclusion

By efficiently harvesting and wisely using finished compost, you can create a self-sustaining garden that thrives from the nutrients meticulously crafted through your composting efforts.

This process not only enhances soil health but also contributes to a more sustainable lifestyle, recycling organic waste into valuable garden resources. As you embrace this cycle, you'll find your garden not just growing plants, but flourishing with life.

With this knowledge in hand, you're well-prepared to apply your finished compost in ways that will enrich your garden like never before.

Happy composting, and may your garden thrive with the fruits of your labor!.

Chapter 16: Compost Tea: Making Liquid Gold for Your Garden

In the rich ecosystem of garden care, compost tea is often heralded as "liquid gold. " This potent brew is a nutrient-rich solution made by steeping compost in water, then applied to plants as a foliar spray or soil drench.

It's a natural booster, packed with beneficial microorganisms, enhancing plant growth and strengthening resilience against pests and diseases. This chapter will guide you step-by-step on creating and using compost tea, ensuring your garden reaps the maximum benefits.

Understanding Compost Tea

To be honest, Compost tea is essentially a liquid form of compost that captures both nutrients and beneficial microbial life.

Unlike traditional compost, which is applied in solid form and breaks down slowly, compost tea provides an immediate nutrient boost. The liquid form is easily absorbed through roots and leaves, providing a quick shot of essential minerals and microbes.

Benefits of Compost Tea

1. **Nutrient Boost:** Compost tea contains a broad spectrum of nutrients that plants need for optimum growth, including nitrogen, phosphorus, potassium, and trace elements.

To be honest, 2. **Microbial Life:** The brewing process encourages the growth of beneficial microorganisms that help suppress plant diseases and promote healthy soil life.

3. **Improved Soil Structure:** Regular application of compost tea improves soil structure, helping it retain moisture and nutrients more effectively.

4. **Increased Plant Vigor:** Plants treated with compost tea often show enhanced growth, increased flowering, and improved fruiting.

5. **Pest and Disease Resistance:** The microbial diversity in compost tea strengthens plant defenses, reducing reliance on harmful chemical pesticides.

Selecting the Right Compost

The foundation of great compost tea is high-quality compost.

Homemade or commercial compost can be used, provided it's mature and pathogen-free. A good compost should be well-aerated, rich in organic material, and filled with life—look for the presence of beneficial fungi, bacteria, and other microorganisms.

Characteristics of Ideal Compost:

- **Texture and Smell:** The compost should be dark and crumbly with an earthy aroma.

Interestingly enough, - **Diversity:** It should consist of mixed materials (e.g., straw, leaves, grass clippings) to ensure a variety of nutrients.

- **Free from contaminants:** Avoid compost that smells rotten or contains non-organic waste.

Tools and Equipment

To make compost tea, you'll need several basic items:

- **5-gallon bucket:** Standard size for small to medium gardens.

- **Aeration device:** An aquarium pump or air stone works well.

At first glance, - **Water:** Use chlorine-free water or let tap water sit for 24 hours to allow chlorine to dissipate.

- **Mesh bag or cheesecloth:** To place your compost in, allowing for easy removal.

- **Molasses or other sugars:** Optional, but provides an additional food source for microbes, promoting faster growth.

Brewing Compost Tea

Step 1: Fill the Bucket

Begin by filling the bucket about halfway with compost.

The quality here is crucial; more isn't always better. A dense solution might not aerate sufficiently and could turn anaerobic, promoting harmful bacteria.

Step 2: Add Water

As strange as it sounds, Fill the bucket with water, leaving a few inches at the top to prevent overflow when aerating.

Remember, the water should be free of chemicals like chlorine that could inhibit microbial growth.

Step 3: Aerate the Solution

Submerge your aeration device in the water and start it.

This will ensure oxygen is constantly mixed into your solution, preventing anaerobic conditions and encouraging beneficial bacterial growth.

Step 4: Add Sugar Source

Stir in a few tablespoons of unsulfured molasses or another sugar source to provide a food boost for the microorganisms.

Step 5: Brew Time

In real life, Let your solution brew for 24 to 48 hours.

The key is keeping the mixture aerobic, so ensure the pump runs continuously. You'll notice foam and bubbles on the surface, indicating active microbial life.

Application and Usage

Applying Compost Tea:

In real life, - **Foliar Spray:** Use a garden sprayer to apply directly to leaves.

This method allows plants to absorb nutrients and microorganisms through their stomata (tiny openings on leaves).

- **Soil Drench:** Pour the tea directly onto the soil around plant bases. This method enhances soil organisms and provides nutrients through the root zone.

Usage Timing:

For best results, apply compost tea early in the morning or late in the afternoon when stomata are open, minimizing evaporation and maximizing absorption.

Application can be done every 14-21 days during the growing season.

Possible Pitfalls and Considerations

While compost tea is incredibly beneficial, certain pitfalls can reduce its effectiveness:

- **Anaerobic Conditions:** An unpleasant smell in your tea indicates anaerobic conditions.

If this happens, discard the batch and ensure proper aeration in your next attempt.

- **Over-fertilization:** Don't let the term "liquid gold" mislead you into overusing compost tea.

Nutrient burn can occur if applied in excessively high concentrations or too frequently.

- **Pathogen Risks:** Always ensure compost is pathogen-free as not to infect plants.

Home-brewed compost tea may not achieve the pathogen control level of commercial products.

Wrap Up

Embracing compost tea as part of your gardening routine offers a sustainable way to boost plant health naturally. By offering nutrients directly available to leafy greens or flowering blossoms, this liquid marvel enhances not only the growth rate but also the resilience and overall vigor of your plants.

Mastering the art of compost tea may take a bit of practice, but the reward—a garden that thrives with life—is well worth the effort. Turn your waste into wealth and let this liquid gold infuse your plants with the life they deserve.

Chapter 17: Integrating Composting into Your Lifestyle

Incorporating composting into your daily routine may seem like a daunting task at first glance, but with a few practical strategies and a little dedication, it can become an effortless part of your life.

By making composting a habitual practice, you can minimize waste, support environmental sustainability, and transform how you connect with your surroundings. This chapter will guide you through seamlessly integrating composting into your everyday lifestyle, ensuring it becomes as natural as taking out the trash or recycling.

Understanding Your Waste

As strange as it sounds, The first step to integrating composting into your lifestyle is understanding the types of waste you generate. Take a week to observe your household's waste habits.

Keep a notebook handy, or use a note-taking app on your phone, to categorize the trash you produce. You'll likely find that a significant portion, particularly food scraps and yard waste, is compostable.

When grocery shopping, consider adopting a mindful approach to reduce the overall volume of waste generated from your kitchen.

Plan meals to minimize leftovers and specialize in purchasing items with minimal packaging. Reducing waste at the source is an excellent step towards easing your entry into home composting.

Setting Up a Convenient System

As strange as it sounds, To make composting second nature, the first thing you need is a convenient system. A countertop compost bin can be a great addition to your kitchen.

These small, manageable containers can hold several days' worth of food scraps. Choose a model with a charcoal filter to keep odors at bay, and make sure it's within easy reach while you cook and clean.

In addition to a kitchen compost container, establish an outdoor compost structure like a compost pile, bin, or tumbler.

Make sure the location is accessible, yet not too close to your living spaces to avoid unwanted smells and pests. Consider weather changes too; placing your compost pile in a sheltered spot can help regulate moisture and temperature, improving the rate of decomposition.

Family and Household Involvement

Get everyone in your household involved.

Educate family members about the benefits of composting and demonstrate how they can contribute. Children can be encouraged to participate through small tasks, like emptying the kitchen container into the outdoor bin or helping to turn the compost pile.

Adults should also share these responsibilities.

Assign weekly or bi-weekly roles on a rotating schedule to ensure everyone plays a part. By involving the whole household, you transform composting from a chore into a shared endeavor, fixing it firmly into your collective routine.

Efficient Sorting Practices

To improve the efficiency of your composting efforts, develop a robust system for sorting waste. Kitchen scraps can be divided into green (nitrogen-rich) and brown (carbon-rich) materials.

Greens include fruits, vegetables, coffee grounds, and fresh grass clippings. Browns include dead leaves, cardboard, and shredded paper. Keeping a balance between these types will accelerate decomposition.

Label containers within your home for compostables, recyclables, and trash to simplify sorting.

Post a list nearby of what can and cannot be composted to minimize mistakes. This setup not only assists individuals in correctly sorting waste but also streamlines the process for the entire household.

Making It a Habit

To be honest, Integrating composting into your lifestyle hinges on making it habitual.

Establish routines: scrape plates directly into the compost bin after meals, empty the kitchen container regularly, and turn the compost pile weekly.

Set reminders on your calendar or phone until these tasks become a natural part of your routine.

Reinforcing these actions will gradually convert them into automatic habits, reducing the mental effort involved.

Dealing with Challenges

It's worth noting that Every new habit faces occasional hurdles.

If odors become an issue, ensure you're maintaining the right balance between greens and browns and aerating the compost regularly. For pest control, make sure your compost bin is sealed and properly layered.

Weather can also present challenges.

In cold climates, consider insulating your compost bin during winter months to sustain microbial activity. In contrast, during hot, dry periods, you may need to add water to keep the materials damp and decomposition ongoing.

Harvesting and Using Your Compost

Many people don't realize this, but Finally, integrate the use of your compost into your gardening practices.

Once the compost is ready, spread it in your garden beds, around trees, or mix it with potting soil for houseplants. Witness the benefits of your efforts as your plants grow healthier and more vibrant.

Invite household members to participate in this rewarding stage, reinforcing the positive outcomes of composting.

As you reap the benefits, share excess compost with neighbors or community gardens, potentially inspiring others to adopt composting themselves.

Staying Committed to Sustainability

Interestingly enough, The journey to a sustainable lifestyle is ongoing, with composting playing a vital role.

Stay informed about community composting initiatives or educational sessions to expand your knowledge and commitment.

Engage with online forums or local groups to share experiences and learn new techniques.

By connecting with fellow composters, you'll stay motivated and open to continuous improvement, ensuring composting remains an effortless and rewarding part of your life.

Incorporating composting into your daily routine requires thoughtful planning, but the ecological and personal benefits are well worth the effort.

By gradually embedding composting into your lifestyle, you foster a proactive approach to sustainability, minimize waste, and enrich the earth—all while cultivating a fulfilling, environmentally-conscious way of life.

Chapter 18: Composting in Small Spaces: Solutions for Urban Dwellers

The hustle and bustle of city life doesn't just mean vibrant culture and endless amenities; it often means limited space, especially when it comes to gardening and composting.

While it's easy for rural and suburban dwellers to toss kitchen scraps into a backyard compost pile, urban living often requires a bit more creativity. This chapter explores practical composting solutions tailored for urban dwellers who are keen to recycle waste and enrich their small gardens or potted plants with nutrient-rich compost.

Understanding the Urban Composting Challenge

To be honest, Living in a densely populated area typically means limited outdoor space, which can pose a unique set of challenges for composting.

However, with innovation and the right approach, you can transform even the tiniest of spaces into an efficient composting station. Whether you live in an apartment, a condo, or a compact urban house, there are numerous ways to make composting work for you.

The key constraints are space, odour control, and managing the right conditions for your compost to process effectively.

But before diving into specific solutions, it's important to understand that urban composting requires a bit of foresight and planning. It invites you to rethink how you handle waste, transforming what was once thrown away into a resource for your plants.

Selecting the Right Composting Method

When it comes to urban composting, size and efficiency are paramount.

Here are a few popular methods that fit well with limited space and ensure a seamless workflow from waste accumulation to compost production:.

1. **Indoor Worm Bins (Vermicomposting)**

Vermicomposting, or worm composting, is one of the most efficient ways to compost in tight spaces.

You'll need a worm bin, red wiggler worms (known for their voracious appetites), and a commitment to maintaining optimal conditions for your little composters.

Setup: Acquire a plastic bin or a specialized worm bin with adequate ventilation.

Drill small air holes for oxygen flow but not big enough for your worms to escape. Line the bin with moist bedding materials such as shredded newspaper or cardboard.

Operation: Add your organic waste (vegetable peels, coffee grounds) in moderation to the bin.

Be mindful of your worm-to-waste ratio to avoid overloading the bin. Worms consume up to half their weight in waste each day, so adjust your inputs accordingly.

Maintenance: Keep the bin in a dark, cool place (between 55-77°F or 13-25°C), and maintain a moisture level akin to a damp sponge.

Turn or fluff the bedding occasionally, and harvest the casting every few months for your plants.

2. **Bokashi Composting**

Originating from Japan, Bokashi composting uses anaerobic fermentation to break down organic matter, allowing it to be done indoors efficiently.

Setup: Obtain a Bokashi bucket, which comes equipped with an airtight lid and a tap for draining excess liquids.

You'll also need Bokashi bran, which contains beneficial microbes crucial for fermentation.

In real life, **Operation:** Layer kitchen waste with Bokashi bran inside the bucket.

Each time you add waste, sprinkle a generous amount of bran, press the contents to remove air pockets, and seal the bucket tightly.

Maintenance: The absence of air prevents odor; however, drain the liquid regularly using the tap.

This by-product, often called "Bokashi tea," can be diluted and used as a plant fertilizer. After two weeks, waste in the bucket can be buried or placed in another container to further break down.

3. **Compact Tumbler Bins**

Let's not forget, If you have a small balcony or rooftop space, a compact tumbler offers neat, contained composting.

Setup: A small, rotating barrel or tumbler allows you easy access and encourages regular aeration by tumbling the contents.

Operation: Add diversified waste materials in layers, balancing green (nitrogen-rich) and brown (carbon-rich) materials. Avoid overloading the drum to ensure space for aeration.

I've seen this happen before: **Maintenance:** Turn the tumbler every few days to mix the contents, accelerating the decomposition.

Keep an eye on the moisture level, adding dry materials if it gets too wet.

Tips for Successful Urban Composting

1.

Balance Your Greens and Browns: Maintaining a healthy ratio of nitrogen-rich green materials (like fruit and vegetable scraps) and carbon-rich brown materials (like shredded paper) is essential for efficient composting.

2.

Odor Management: Effective method selection is key, as improperly managed compost can attract pests or produce unpleasant odors. Regularly turning and aerating your compost, as well as covering fresh waste, will help.

3. **Stay Consistent:** Regular attention and maintenance ensure your composting system operates smoothly. Add waste regularly, monitor conditions, and harvest compost when ready.

4.

Educate Your Household: A successful urban compost system requires cooperation from all members of your household, encouraging them to contribute to the compost with care.

Let's not forget, 5.

Leverage Community Resources: Consider reaching out to urban farming collectives or community gardens that might accept overflow or contribute guidance to your composting journey.

Urban composting can significantly reduce household waste, lowering your carbon footprint while nurturing your garden's soil microbiome. As an urban dweller, overcoming space and resource limitations to compost effectively is a gratifying task, fostering a sense of sustainability and responsibility.

Remember, minimal space doesn't equate to minimal impact. Your efforts in composting can contribute to a healthier planet and more fertile gardening landscape. With the right tools and techniques, every urban environment can become a hub of compost production, closing the loop on waste and growth in an inspiring act of urban resilience and innovation.

Chapter 19: Community Composting: Engaging with Local Initiatives

Engaging with community composting initiatives is a rewarding way to help manage local waste, feed gardens, and strengthen community ties.

Community composting serves as an efficient way to include those who may not have the space or resources to compost at home, and it offers learning opportunities for those interested in sustainable practices. In this chapter, we'll explore the practical aspects of community composting, from understanding how these initiatives work to tips on starting and sustaining one in your neighborhood.

Understanding Community Composting

At its core, community composting involves the collective effort of a neighborhood or group to manage organic waste and produce compost.

These initiatives can be as simple as a shared compost bin in a community garden or as organized as a municipal collection scheme with drop-off points and volunteer management teams. The main goal is to divert organic waste from landfills, primarily food scraps and yard waste, and turn it into fertilizer that can nourish community green spaces.

Benefits of Community Composting

Community composting not only helps in reducing landfill waste, but it also educates people about waste management, builds community spirit, and can even save money for participants on waste disposal fees.

For local authorities, it offers a cost-effective solution to waste management challenges by involving the community in sustainability initiatives. Additionally, participants can gain access to nutrient-rich compost for their gardens, improving soil health without the need for synthetic fertilizers.

Types of Community Composting

1.

Drop-Off Stations: Designed for convenience, these stations allow community members to deposit their organic waste in designated areas. Collected materials are then processed into compost by volunteers or a central facility.

2.

Shared Compost Bins: Usually established in community gardens, these bins allow members to contribute organic waste and participate in the maintenance of the compost, ensuring it's turned and balanced correctly.

At first glance, 3.

Municipal Compost Programs: Run by the local government, these programs often include curbside pickup of organic waste or strategic placement of compost bins across neighborhoods. They usually handle larger volumes and require formal structure and regulations.

4.

Cooperative Composting: Here, several households join forces to share compost bins in a volunteer-managed setting. This approach might involve rotating responsibilities for monitoring and turning the compost.

Getting Involved with Local Initiatives

Believe it or not, To engage with community composting, start by researching existing programs in your area.

You can do this by contacting local environmental organizations, checking municipal websites, or visiting community centers for information. Participation might entail regular contributions of your kitchen scraps, volunteering for compost site maintenance, or even assisting with educational outreach.

Tips for Participation

- **Understand Guidelines:** Adhere to what items can be composted within the initiative.

Some community programs have specific requirements, such as excluding meat or dairy products, to avoid odors and attract pests.

As strange as it sounds, - **Contribute Regularly:** Consistent participation ensures that the composting process has a steady input of materials and helps maintain the momentum of the program.

- **Join a Volunteer Group:** Many community composting initiatives require hands-on help, from turning the pile to organizing workshops.

Volunteering also offers an opportunity to learn more about composting techniques and meet like-minded individuals.

- **Educate Others:** As you become familiar with composting, consider sharing your knowledge with neighbors or through community workshops.

Education is key to the expansion and success of community composting.

Establishing a Community Composting Initiative

If no local programs exist, consider spearheading a community composting initiative. Here's a step-by-step guide to get started:

Step 1: Garner Interest

Gauge interest by talking to your neighbors, local schools, or community centers.

Organize an initial meeting to gather ideas and assess the level of commitment. A strong community interest will be essential for longevity.

Step 2: Choose a Location

Find a location that is accessible, central, and has enough space for a composting area.

Potential sites might include community gardens, parks, or a designated vacant lot. Permission may be needed from property owners or local authorities.

Step 3: Set Up Composting Systems

Decide on the type of compost system that fits your community's needs and size.

You will need appropriate bins or piles, aeration tools (like pitchforks), and clear signage indicating what can and cannot be composted. Consider weatherproofing with tarps or lids.

Step 4: Create Guidelines and Rotations

Interestingly enough, Develop simple, clear guidelines for participants about what types of organic material are acceptable, how to contribute, and how to help maintain the system.

If needed, organize rotations for compost maintenance tasks like turning and watering the piles.

Step 5: Secure Funding and Resources

Funding can be sourced from local government grants, environmental non-profits, or community fundraisers.

Resources, including bins and tools, might be donated by local businesses or purchased with grant money.

Step 6: Promote and Educate

Use local networks, social media, newsletters, and flyers to promote the initiative.

Hosting regular workshops or information sessions will also help educate the community about the benefits and processes of composting.

Step 7: Monitor and Evaluate

Interestingly enough, Regularly evaluate the progress of the initiative.

Are the bins being used correctly? Is there a need for more space? Gathering feedback will help to adapt and improve processes.

Overcoming Challenges

While community composting is beneficial, it's not without its challenges. Common issues include mismanagement of waste, contamination from improper materials, and sometimes skepticism from community members.

The key to overcoming these challenges is communication and education. Providing clear guidelines and continuous outreach can ensure participants know how to properly partake in the composting efforts.

To be honest, Finally, fostering a sense of ownership among participants can mitigate issues.

When community members feel responsible and invested, they are more likely to adhere to guidelines and help inspire others to participate.

Incorporating community composting into your neighborhood can have profound environmental and social impacts.

As you engage with these local initiatives, you not only contribute to a healthier planet but also cultivate a stronger, more sustainable community.

Chapter 20: The Environmental Impact of Composting: Contributing to a Sustainable Future

In recent years, environmental sustainability has grown beyond academic discourse to become a personal and collective commitment towards a healthier planet.

Composting, despite its humble and earthy nature, plays a pivotal role in this global effort. As the world grapples with challenges like climate change, waste management, and soil degradation, composting emerges as a simple yet effective solution that promises profound benefits.

Reducing Waste and Landfill Burden

Believe it or not, The modern waste management crisis is an urgent environmental issue, characterized by overflowing landfills and the unsustainable consumption of resources.

According to the Environmental Protection Agency (EPA), approximately 30% of what we dispose of is organic waste, which can be composted rather than thrown away. By integrating composting into everyday practices, individuals have the power to substantially reduce the volume of waste sent to overcrowded landfills.

When organic materials decompose anaerobically in landfills, they produce methane, a potent greenhouse gas that contributes significantly to global warming.

Composting, on the other hand, is an aerobic process, converting organic waste into a valuable resource while minimizing methane emissions. As more households adopt composting, the cumulative reduction in methane emissions could play a crucial role in mitigating climate change.

Enhancing Soil Health and Productivity

Interestingly enough, Soil health is fundamental to a sustainable food system and ecosystem balance.

However, intensive agricultural practices deplete soil nutrients and structure, leading to erosion, reduced fertility, and desertification. Composting restores these vital nutrients back into the earth, revitalizing soil health and increasing its capacity to retain moisture and support plant life.

Composted material enhances the soil's ability to bind together, reducing erosion and water runoff.

This not only conserves precious water resources but also helps prevent harmful substances from leaching into nearby water bodies. By improving soil fertility and structure, composting supports plant growth, promotes biodiversity, and contributes to a more resilient agricultural system.

Reducing Dependence on Chemical Fertilizers

Let's not forget, Modern agriculture's heavy dependence on chemical fertilizers poses environmental and health risks, including nutrient runoff, waterway contamination, and soil degradation.

Composting offers an organic alternative that fosters sustainable farming and gardening practices. The nutrients released during compost decomposition are absorbed by plants gradually, providing a steady supply of essential nutrients without the risk of over-fertilization.

Moreover, composting reduces the demand for synthetic fertilizers, decreasing their production and application, both of which contribute significantly to greenhouse gas emissions.

By embracing composting, gardeners and farmers can shift towards more regenerative practices that preserve ecosystem health and reduce their environmental footprint.

Supporting Biodiversity

I've seen this happen before: Composting fosters a rich teeming ecosystem, from the smallest microorganisms to larger soil fauna, creating a haven for biodiversity.

These organisms, decomposers by nature, play an essential role in the waste-to-compost transformation process. They break down organic matter, releasing nutrients and helping maintain the delicate balance of our ecosystems.

A nutrient-rich soil enhanced by compost is more hospitable to plant life, promoting diverse plant growth.

In turn, healthy plants provide habitats and food for insects, birds, and other wildlife. This biodiversity is integral to maintaining resilient ecosystems that can withstand changes and pressures including those induced by climate change.

Community Empowerment and Environmental Education

On a broader scale, composting serves as an educational tool and community-building initiative.

Community composting programs not only divert waste from landfills but also educate participants about sustainable practices and environmental stewardship. They engage local populations in hands-on activities that highlight the importance of organic recycling, closing the loop between consumption and waste.

Educational initiatives can inspire a new generation to adopt eco-friendly lifestyles.

Schools and community centers can create demonstrable compost piles, teaching sustainability and responsibility through active participation. By working together, communities can build resilience and drive collective action toward sustainable development.

Contributing to a Circular Economy

Believe it or not, The notion of a circular economy revolves around reducing waste, extending the lifecycle of materials, and regenerating natural systems.

Composting embodies these principles by transforming waste into vital nutrients for new growth. It completes the cycle of consumption and replenishment, creating a circular flow of resources that reduces the need for raw materials and lessens environmental stresses.

Incorporating composting into daily life implies more conscientious consumption choices and encourages mindfulness in generating waste.

By taking organic waste out of the linear waste stream, individuals actively partake in a regenerative cycle that reinforces environmental sustainability.

Practical Steps to Enhance Composting Impact

I've seen this happen before: To maximize the environmental benefits of composting, consider these practical steps:

1. **Spread Awareness**: Share your knowledge about composting with friends, family, and neighbors. Encouraging others to join in the effort amplifies its positive environmental impact.

2. **Create Community Initiatives**: For those who may not have the space to compost individually, community composting programs provide an accessible, collective solution.

3.

Support Local Policies: Advocate for municipal and state policies that promote composting efforts and infrastructure. Public compost bins, incentives, and educational campaigns can enhance composting rates in urban and suburban settings.

4. **Educate Yourself and Others**: Keep informed about the latest composting techniques and practices. Host workshops or attend local events to share strategies and resources.

5. **Practice Backyard Composting**: Start or improve your compost bin or pile to reduce personal waste footprints and nourish your garden naturally.

6. **Celebrity Endorsements**: Secure endorsements from local celebrities or influencers to give a wider reach to the practice.

By integrating these steps into our routine practices, composters around the world can contribute to a sustainable future, fostering environmental consciousness and enriching the natural world.

Composting is not merely about managing waste; it's about investing in a future where ecological balance is restored, and sustainable living becomes a global norm.

As individuals take ownership of their environmental impact, composting stands as a testament to the powerful changes that can emerge from simple, daily actions. In practicing composting, we hold in our hands the potential to nourish both our gardens and our planet, one piece of decomposing organic matter at a time.

About the Author

T. S. AVINI is a writer who believes in the power of sharing life's lessons. Having faced hardship and pain, he turned to writing as a form of self-expression and healing. Today, he pours his heart into every book, aiming to help others navigate their own struggles and find strength in vulnerability.